Grass-Based Dairy Farming

15 farmers share considerations
for starting your own
grass-based dairy

Cover photo: Doyle Yoder
Book and cover design: Amy Wengerd

ISBN 10: 1-933753-09-9
ISBN 13: 978-1-933753-09-6

Carlisle Press
WALNUT CREEK

2701 T.R. 421
Sugarcreek, OH 44681
1.800.852.4482

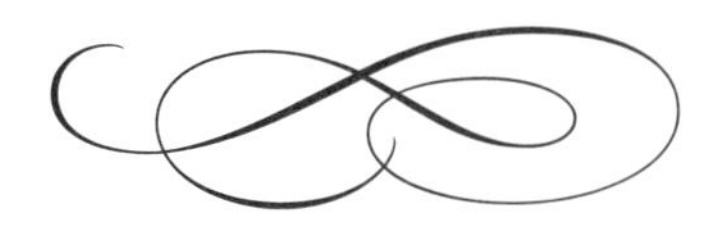

Preface

This book has its birth in the barns and fields of farmers who were willing and determined to restore farming to a prominent position as the occupation of choice for Amish young people. In doing so, they have sifted through many traditional practices and blown out the chaff. They have perked the ears of many of our young people who have a marginal interest in farming. They have also raised a lot of questions. With this book they have anticipated and answered some of them. Enough, they hope, to stir more interest and at the same time place a corner marker on the boundaries of reality.

The fifteen writers have chosen anonymity but we welcome your responses. Send to: Grass-Based Dairy Farming, 2673 TR 421, Sugarcreek, OH 44681.

To distill the concept of rotational grass-based dairy farming into a few simple sentences that are easily understood by all may be difficult. It may even be counter-productive if not balanced with the twin forces of reality and practicality. Nevertheless consider the concept of rotational grass-based farming thus:

Grass-Based Dairy Farming— The Concept

We, in the spirit of good land stewardship, are managers of the primary plant of God's creation—grass. We harvest this grass in the most ecologically correct manner we can by our cooperation with the laws of nature and dictates of the bovine species. We then sell her milk as a reward for our stewardship and gain the satisfaction of providing for our families and our communities in a manner that violates neither the earth nor those that tread upon it.

Sustainable Farms

•Profitable enough to cash flow now and also generates enough capital to allow the next generation to pay a fair price for the farm

• Fits our way of life, our values and our goals

• Can be done by most farmers, not just the best

• Increases rather than depletes the farm's resources

• Attracts our children to the farming way of life

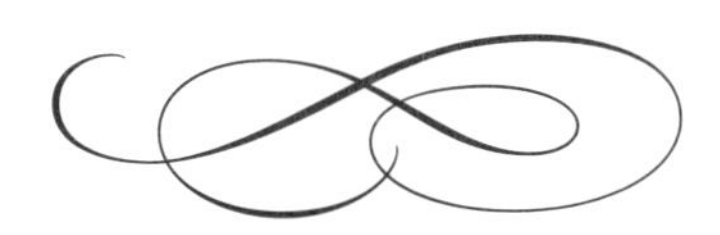

Table of Contents

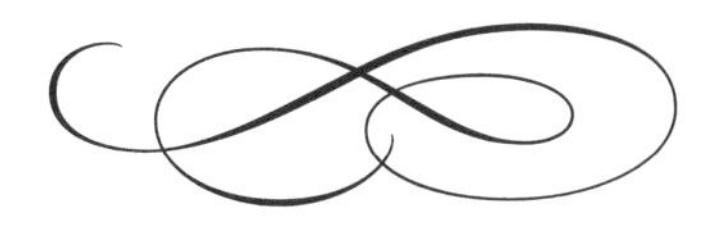

Farming Vision

"Where there is no vision the people perish" (Proverbs 29:18). What is a vision of farming? It is an idea of the advantages and the methods needed to accomplish such. These have to be compatible with and beneficial to our core values and beliefs. In this case, for most of us, favorable for remaining Amish.

Farming: Most Basic Occupation

Farming is the most basic and essential of all livelihoods. "To dress and to keep" the land and "to have dominion" over domestic animals was part of the first commandment to mankind. Farming is the most necessary occupation to provide nourishment for all mankind. One is a steward of the natural elements of creation, and it can be very satisfying, even at times exciting, to be a co-laborer in the chain of food for life.

It is more than a mere livelihood for family income. If befittingly done it is an ideal place to raise a family. One can be in a rural network of interdependence with Christian neighbors where solid Christian values are nourished and flourish. It is a way of life. It is almost impossible to find a replacement for it that will serve us as this has for 300 years. A simple farming culture brings a greater awareness of working with creation and a patient waiting of the "early and latter rains." It creates a sense of commonness by working with soil, plants, and animals, all in a cohesive family unit. It accords so well with a plain faith community.

For this reason we feel as much should be done as possible to have farming be such an opportunity. A concentrated effort and looking into it is needed. We think, for the Amish, there is still a good, even great, possibility for such. But for most of us we probably will have to change some of our thinking and methods to succeed.

Our Background

Back in Europe, our forefathers in Alsac-Lorraine (France) and southern Germany, after fleeing from Switzerland, slowly carved out a reputation of being good, careful stewards of the land. While still oppressed because of their religion, they were tolerated only on lean, war-ravaged land at first. They willingly lived in a more remote countryside, where they were allowed freedom of religious beliefs and practices.

Previous native farmers had farmed out the soil's fertility mostly by growing continuous grain, followed by a fallow year. In addition to being war-ravaged, the land now had poor fertility. The agricultural economy was bleak. To this land came determined religious refugees and were allowed to stay.

They refused the easier atmosphere of commerce and industry that they could have enjoyed farther north.

These rugged Anabaptist farmers became pioneers in growing legumes and grasses together in spring, summer, and fall meadows. They learned that sweetening the fallow soil with limestone and gypsum helped build fertility. As Swiss people, they were used to milking cows and making cheese seasonally. This brought in much-needed animal manure. They switched from the sickle to the scythe for harvesting hay and small grain. They rotated crops and pastured animals. In winter they now kept animals, which previous natives had not done. This manure, partially composted, was spread in spring. All this combined for a gradual noticeable difference in the land. Only after many years did the government reluctantly admit that these simple farmers had skills their own natives lacked. They sent out so-called agricultural experts to determine what methods brought about this increased sustainability.

It was the combination of bovine animals, legumes, grasses, limestone, and manure. The switch from sickle to scythe allowed for more winter feeding of hay and small grain. Barns were built and more animals and feed could be kept, making for more manure, and gradually the rural economy became better. The organic matter in the soil increased and erosion became less and water retention better. There were more people on the *farmhofs*, and the refugees became known for their farming skills.

Grass and Ignoring Grass

In 1937 *Hoard's Dairyman* still had these words in their Annual Farm Handbook: "We cannot conceive of a cheaper and easier way to produce milk than from pasture."

This is also the claim of our work in this book. In the creation account, grass is the first-mentioned plant, and it is by far the most numerous and common plant on earth. It grows and regrows to make it, if properly managed, the overall most valuable plant a dairy farmer has.

The last 60-70 years grass became increasingly ignored by modern-day farming. The cry was for more corn, more beans, and more alfalfa. The reason for this shift in focus was connected to modern tractor farming, chemicals, fertilizer, confinement, dairying, and genetics. Extension people and agribusiness sales people all helped along...and filled their own pockets.

Sure, the cows gave much more milk, but the cost of producing a hundredweight went up and up and up till today the US has one of the highest cost per hundredweight of the major milk-producing countries of the world. Thanks largely to ignoring grazing pastures as the main feed.

In 1950 over 50% of all farmland between the East Coast and the Rocky Mountains was still in grass or hay. But due to ignoring grass, today it is probably less than 25%. It is almost no wonder that in the last 50 years, on average, 200 farmers have quit every week!

This is the background for the return to grass-based farming. Of course, many things changed in the last 50-60 years—land prices, taxes, building supplies, living costs, and medical expenses, which we have almost no ability to change.

Grass-based farming is not new, neither is it a full return to pre-modern farming. It is, however, a return to using pasture as our main feed. We let the cows harvest grass for almost no cost, and in return get manure spread, have drastically less vet costs, and have more heifers to sell because of better longevity.

Financial Soundness

We do not want to paint such a rosy picture that this persuades you that one simply can't go wrong if he gets into grazing. Don't get the idea that to succeed you have only to be a grazier or to become an organic grazier. It simply is not that easy!

This is the greater underlying concern of this booklet. Your operational costs and management have to be efficient and sound. Beginning graziers need a vision, but they also need a vision to get sound advice from other more experienced grass-based farmers and the people they borrow money from.

This grass-based movement has opened more sharing of actual financial figures among farmers than ever before. Due to sometimes overoptimism, these figures need to be carefully checked for accuracy prior to using them for income projection.

When a novice grazier needs to spend more than $30,000 to $50,000 for starting up, he should get advice from experienced and knowledgeable graziers. "In the multitude of counselors there is safety."

In almost every new movement similar to this there is often a lag of practical experience that only more time will bring. This grazing movement is no exception.

Our Amish school movement, starting around 1950, was just like that at first. This grass movement can be similar if we are open enough for a different way. We need to keep it simple and financially sound. And certainly within the limits of being Amish. If not, there is no gain in us trying to promote it.

Reviving a Faded Vision

In the last two to three generations, farm families have been diminishing. The narrow-mindedness of specialized American agriculture has so often narrowed farming down to only one enterprise. The result is that often only one or two people are needed on the farm. All others can become non-farm employees. So the farming families depopulate themselves.

Our vision should include repopulating our farms with more people. With better cash flow and returns from grass-based dairies possible, we would love seeing dairy as the main enterprise, but having other smaller projects supplementing the larger one. These need to be flexible enough to grow as the family's help grows, and also be able to roll back some again as the help leaves when the children leave on their own.

The loss of profitability with hogs and hens has been a large factor in this. But today there are opportunities with hogs and hens on a grass-based farm, plus produce, greenhouses, small fruits, and seasonal shops.

This will take some stretching of our vision to accomplish. We do not see a 50-cow grass-based dairy handled by mostly one person as the sum of our vision. Rather we see the family that can work together daily on a multi-project farm and truly be a farm family.

Conclusion

Great potential lies in grass-based farming. A concentrated look into it is justified. Grass-based farming can be adapted well to more conservative setups and methods than often described in this book.

We need to be honest and also look for snares. All such newer movements also have negatives which require modification. It is wise to acknowledge that and try to avoid such.

But the overall advantage of retaining and gaining farmers, if befittingly done, is so great it is not wise to ignore the potential this type farming has for us.

A fulfilled vision comes only when we make sound decisions, work hard, and have persistence.

"Write the vision and make it plain" (Habakkuk 2:2).

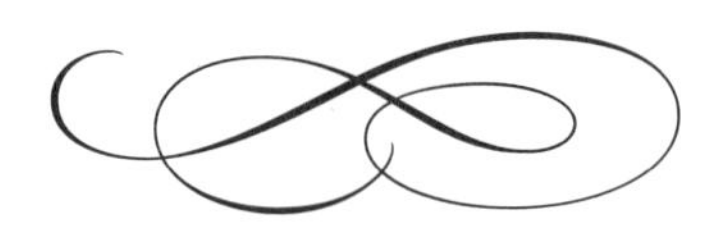

Understanding Rotational Grazing

The rather recent emergence of grass-based farming is again giving hope that our farming must not be lost. In some communities it has already sparked a return to farming.

What it does is provide a cheaper way of producing milk and operating a farm, and helps create a better profit for the dairy. The simple, learned-by-experience skill of making most of the milk from green growing grass makes the entire dairy business an even playing field for the small farmer. In fact, I feel it is an actual advantage!

Instead of high production with large, expensive cows, and high elevator, vet, fertilizer, and chemical bills, the skill is to harvest as much milk from grass by the cow as possible. There is no cheaper milk made than from the cow on lush grass. The skill to be learned is how to have this lush grass growing longer than only during the month of May.

A point of the same importance to us is that this can be done in a way that fits into our way of life.

Equally important is that the average among us can do it,

providing we apply ourselves, are willing to learn, and are persistent in applying the rules of rotational grazing. What it takes is mostly the cow and grass, and a boss. From haymaking, corn growing, and manure hauling, it does not take more equipment. Most times less.

The more and longer the cow can harvest good grass herself, the more profitable. It's the opposite of the conventional way which so often just takes more machinery to make more and better feed.

In some ways, it is farming more like they did before all the tractors, chemicals, antibiotics, fertilizers, confinement buildings, and concrete came along. Before WWII a high percentage of herds were mostly seasonal or all seasonal. They took advantage of having the cows dry when the feed cost to make milk was highest—in winter. It costs at least twice as much to make milk in winter than when grazing on green grass. Most times a lot more. Yes, even on Amish farms! Once we learned how to graze many more nutrients per acre, we wondered how it happened that we missed all this for so long.

When these methods are used in good management, farming can again be profitable enough to make a living and pay for the farm.

Earlier Experts

The cow was created for the very purpose of converting grass into milk. In grass-based farming, the cow and her production shifts to second place as the focus goes more to grass and soil fertility. It's not that production is not important, but grass and soil fertility are just more important.

This is the key to understanding how rotational grazing dairy farming works. In the long term it is the most possible to attain by the largest group of farmers, especially us Amish farmers, who have such a deep background of dairy farming.

In our search for sustainability, we were amazed how it was promoted up to 1950 by very reliable experts that pasture was the greatest key to profitability.

Hoard's Dairyman 1937

"An acre of good pasture under proper management will produce more pounds of digestible nutrients in a year than the same acreage devoted

to a crop rotation of corn, oats, wheat, and hay.

This feed will be produced at less than 5% of the labor cost. The cost of producing milk on good pasture is only 27% as great as the cost of production of grain and hay.

The amount of animal feed produced on a pasture can be increased 50% by dividing the pasture into three paddocks and rotate graze them.

The amount of feed produced can be increased four-fold if the pasture is divided into seven parts and each field pastured four days and rested twenty-four days.

Erosion losses from land in pasture is only 12% of that from row crops and less than 2% of that where corn is grown continuously."

(And we thought our knowledge was so much more advanced now than theirs was back in those days.)

Rotational grazing is nothing new. Only a forgotten or ignored method. Read on:

"The time to graze any pasture is when growth is new and high in food value. As the grass becomes older the food value decreases.

Get the food value while it is there. Then save the grass plants by keeping all stock off while plants are 'having regrowth.'"

In March 5, 1951, *Farm Journal* reported, *"Some day we'll change pasture every morning!"*

Two years later in 1953, they had an article declaring that "some day" is already here.

"Daily changing pasture is so new only a handful of dairymen are doing it, but with good results. One dairyman is changing every twelve hours with most spectacular results."

They quoted dairymen from New England, California, Minnesota, and Virginia that were very well satisfied with such a practice.

A Dairy Science Book by Peterson, 1939, used in a Dairy Science Ag Course by the University of Minnesota, says *"One of the most important but generally neglected factors in economical milk productions is that of pasture* (page 479). *It is also obvious that under favorable conditions pasture yields may yield nutrients that compare favorably with those from rotation crops at much less expense... On most dairy farms no crop producing land will yield larger returns from intelligent management than will pasture. As a rule pasture is our own neglected crop"* (page 483).

In another part of the book a study compared cost of milk

produced on pastures and cost of milk produced in winter on dry feed of hay and grain. The pasture-produced milk cost was only one-fourth of that produced in winter!

These reports were just on the edge of the dynamics of managed intensive grazing.

But What Happened?

For about the next forty years there was very little, almost no, focus on grass. There was silence. Instead, the focus was on the cow and increasing her production.

The tractor came roaring on the dairy farms after 1950. Artificial oil-based fertilizer became widespread. AI began and a cow was bred for greater corn, corn silage, and soybean consumption. National corporate feed companies flourished. Hybrid corn and alfalfa became dominant crops. Tractors began roaming the pastures, plowing everything under, while more and more the cows stayed at home on concrete. Herbicides, pesticides, fungicides, antibiotics, and hormones became expensive cures for many ills, weeds, and pests.

Almost all the "experts" were claiming "higher production" is the way. Higher production by fossil fuel, synthetics, and genetics. But it became the high road to eventual demise by ten thousands of farmers.

While many good, new, useful products and methods were introduced, they also gradually steered the dairyman to become supplier-bound.

The USDA Farm Policy kept subsidizing the farmers so they could continue farming, they claimed. Who got most of that money? Agribusiness. The farmer became merely the pipeline.

In the 1950s over 50% of all farmland from Iowa to the Atlantic was still in grass or hay. That changed drastically as the Dairy Belt switched from grass and hay to corn and beans.

We who are interested in dairy farming need to think outside the box and start filling our own pockets, and not just fork our money over to agribusiness until we can't make it anymore!

On to Grass Farming Today

Grass farming is having at least the greater part of your farm in good grass. More often than not, it is having all of it in grass. This is harvested by cows (or other animals) at almost no cost.

We'll start with the cow. How does she eat in her natural way? She slides out her tongue on a side of her mouth and grabs a tongue full of grass and tears if off sideways. Slipping it into her mouth, she goes for the next bunch. It's interesting to watch and note important points of grazing.

For the best grazing, the grass must be 8-10" high, a dense stand, soft and lush, clean and green. The tearing off point is at four inches normally. Every swipe of her tongue puts a good-sized tuft of grass into her mouth. This is top milk-producing grass. A cow usually makes only so many bites before she lies down to chew her cud. The better the bites, the better you are feeding her.

For the best grazing, go in at 10" and out at 4", or go in when she can graze with grass tips at her eye level and move her out of the paddock when you can see her nose.

The grass she left is still a plant, and not a stubble, if previously the grass has been 6-8" tall. The leaves remaining on that grazed grass plant will still catch sunshine and start growing again. That way you have the blades helping growth as well as the roots.

In comparison, when you mow for hay the grass or legume is at least 12" or more high, often two feet or more. The last ten days or so of its growth, the sun cannot shine on the bottom four inches because of too much leaf canopy on top. So when you mow it off at 1½-2" you have only mostly brown stubble left. The roots then are the only source of energy to get regrowth until new leaves develop. That is why a managed grazed pasture has quicker regrowth than a mowed hay field—a big advantage in a grazing season! Everything she puts in her mouth is much higher quality than the more fibrous stem of hay.

In early spring, as the grass greens and starts growing, (only 2" high) you can *fast graze* over larger paddocks and still feed stored feed. But do not let them mud up your pasture. A little pugging doesn't matter, but do not let them muddy it up.

Fast grazing is done in the spring and has two purposes. First, it keeps the grass short allowing the warm spring sun to warm the soil faster. (At no other time of the year do we want the soil exposed to the drying out of the sun.) Second, the cows graze short, tender grass high in protein, boosting spring milk production.

Fast grazing is done by giving your cows an entire paddock, consisting of three to five acres depending on herd size and lay of land. As the soil warms and grass grows faster, you will subdivide this paddock with crossfences but for now your cows *fast graze* with this larger paddock for twelve hours at a time in a seven to ten day rotation.

In open grazing the cow is your boss. (Unfortunately she has only instincts but not much intelligence. You are supposed to have that!) What she does the first while is really enjoy herself. She swipes away the best grass first, then the best remaining yet, and again the best remaining. But a week later the first "best" is maybe again 2-3" high and snap! She gets it again but a little shorter this time. The soil is now more apt to dry out with continual short grazing.

That way the good and best grass gets eaten every time. It depletes the root reserves and becomes weaker. The fair and lower-quality grass goes uneaten until it is too stemmy. All this time many plants get tramped upon while "Bossy" is looking for the best remaining grass. Your pasture cannot improve that way. You have to become the boss.

That is why there is such an advantage in controlled rotation grazing. In a paddock you can control what the cow grazes. By having a herd, say twenty cows, on a half acre, they will uniformly graze all the grass much more. In the meantime, all the rest of your grass in the other paddocks is growing unmolested and untrampled.

By uniform grazing and not overgrazing, regrowth is so much better. By uniform grazing there is much more uniform spread of their manure. By better spread of manure there is better spread of microbes in the soil.

All these and more add up to the great advantage of rotation grazing versus open pasture grazing.

Rotation grazing allows the creation of more lush stages grazing like there is in early to mid May. You cannot expect to have May-like grazing all summer and fall, but you can sure help to create that condition more often.

The cows should never be in a paddock longer than three days. Actually, changing at every milking is optimum grazing. In three days some grasses are sprouting new leaves, and to be trampled upon or nibbled at this stage is a loss.

How to Begin

Experiment with how big an area your herd needs to graze it down to four inches in whatever amount of time you want to have them in the paddock. Don't be too exact; just a general height.

Be ready to give them another paddock when they have grazed enough for this time.

Have someone help you the first couple times. The cows will soon catch on that you are giving them new grass and will follow you as you move them to different pastures.

From April 20 to June 20, it is approximately one cow per acre. From June 20 to September 1, roughly two acres. From September 1 on, we stockpile the surplus to graze after the growing season is past, but not the grazing season. This will be November and we've already gone until December before all grazing was over.

We let the cows graze plenty close late in the grazing season and don't leave much residue by December.

But this I say, he which soweth sparingly shall reap also sparingly; and he which soweth bountifully shall reap also bountifully (II Corinthians 9:6).

The Farm Family

The term "family farm" is often used to describe our Amish farming idea. The dictionary describes family thus: father, mother, and children. A family, then, must be more than one person alone. At creation, when God created man, He said, "It is not good that man should be alone." He then made woman and gave her to man for a helpmeet.

Trying to farm without the support and interest of your wife and children becomes very complicated. A supportive wife is a great blessing to her husband. She is there to provide encouragement, prepare food for her family, make and wash clothing, and help with farm chores and bookkeeping. In return, her husband helps and supports his wife in whatever she needs. Ecclesiastes says, "Two are better than one; because they have a good reward for their labour. For if they fall, the one will lift up his fellow; but woe to him that is alone when he falleth; for he hath not another to help him up."

Good communication is always necessary and avoids many problems. We don't plan to fail, but at times we fail to plan. Husband and wife should always plan together on major projects. Get your children involved in the planning stage whenever possible and as their age allows. If you are a farming start-up it is very helpful if your wife can be involved with the farm work. It is helpful if your wife knows how to operate the milking equipment and what certain marks on the cows stand for (e.g. treated cow, fresh cow, bad quarter, dry cow, etc.). She should be well enough acquainted with farm paddocks (by number or name) so that she knows what is being referred to. Try to make sure you leave clear directions for her (write things down!) when you leave the farm and expect a veterinarian or feed truck to arrive in your absence. Care should be taken, however, if she has small children to care for, that you structure farm duties so as to allow for plenty of time and energy to devote to your children. Too much farm work for your wife may lead to neglect of your children which gives a poor return in the long run. Wise is the husband who keeps his thumb on the pulse of his wife when her health allows for less work on the farm. Husbands are instructed to "dwell with them [our wives] according to knowledge" (1 Peter 3:7).

At times things do go wrong. When they do, remember, land is forgiving, grass is forgiving, cows are forgiving (if not abused or ignored), so people need to be forgiving as well. Forgiving, just as God has forgiven us through Jesus Christ. In this way we can move forward and the Lord's blessing can be upon us.

A righteous man regardeth the life of his beast: but the tender mercies of the wicked are cruel (Proverbs 12:10).

Agriculture is the most useful of the occupations of man.

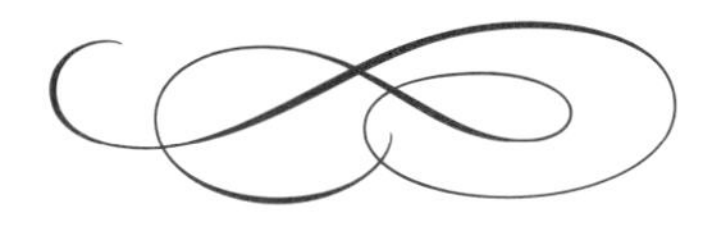

Farm Finances

Financial Planning

There may have been a time when young couples could begin farming without much financial planning. Almost all the Amish farmed the same, land prices were reasonable, and start-up costs relatively low. Sadly, those times are long gone.

Today we are faced with ever higher costs. Land prices are high. Equipment, facilities, and inputs like fuel, feed, etc. keep ratcheting higher. So today it is wise to look before we leap.

Consider yourself blessed if you have the privilege of farming the home place. Not only does this usually simplify finances; there is truly a blessing in a father helping a son (or son-in-law) get started, then in turn the son helping the father in his old age, and hopefully another generation coming on. What a beautiful picture if it can be done in harmony and goodwill!

But even so, what used to work for the father may or may not work for his son. With the ever higher costs come the ever higher number of cows. Facilities that worked fine for 15 cows may no longer be satisfactory for 30 to 50 cows.

What about a young couple buying a farm on their own? Sadly, here in Holmes County those opportunities are few and far between. New communities have been started which offer opportunities for the brave at heart.

Whichever way we begin farming, there will usually be some debt and at times a lot of debt. If you are thinking of buying a farm, now would be a good time to honestly ask yourself: Can I handle the stress of a high debt load? There are few occupations as stressful as dairying in a drought with a high debt load. As I think back, there were times when I wondered if it would rain in time to save us from ruin. But God has always provided a way through.

As we seek our life occupation let's ask our loving heavenly Father for help and guidance. To quote Solomon in Proverbs 3:6: "In all thy ways acknowledge him, and he shall direct thy paths."

So now that you have decided to take the step, how do we plan for it? Start-up costs will be covered elsewhere in this book. Don't be afraid to ask for advice. Is there an acquaintance with an operation similar to what you are planning? Get a good idea on what it will cost for your specific farm.

Rule number one will be to make sure the money is available before you begin. You can borrow from a bank or one of the various lending institutions. Compare interest rates and terms. Is interest fixed or variable? Making payments twice a month instead of once a month reduces interest expense. If you are young, you may need a cosigner before you can get a loan.

Private money may also be a good option. With private money comes increased accountability to the lender (family member or friend). Be careful to have the details of the loan carefully spelled out and written down. Either pay back the loan exactly as you agreed or talk to your lender if you run into difficulty and can't pay on time. Never make him have to interpret your silence. Borrowing from friends and family has great advantages if done correctly. It also carries risk of ruined relationships when poorly structured or administrated.

Rule number two is just as important as number one. Make sure you have a plan in place to address your debts. I assume you are in dairy. Generally, I figure the first 15 or 20 cows will pay for living expenses including medical costs and property taxes. Sure, a young couple living frugally can make do with less. But let's not underestimate our expenses. Medical expenses have become a burden to all societies. The Amish have a good health care plan. The plan is helping each other. Let's not let anybody have an undue burden of medical bills.

So how many cows do we need to address a debt of, let's say, $100,000 at 7% interest? Keep in mind that these figures are derived from average conventional milk prices. So let's figure our net profit at $1,000 per cow per year. Some may say this is conservative. But just as surely as we don't want to underestimate our living expenses, we don't want to overestimate our income. So I would like to see at the very minimum, 10 cows for $100,000 of debt load after the 20 that we need for our living expenses. So that would be 30 cows for $100,000, or 40 for $200,000, and so on.

Once we start figuring interest rates, we can see why it is wise to keep start-up costs at a minimum. For example, buy used equipment instead of new or make do with the facilities that you have for a while. If you can save $50,000 by being conservative, that would save you $3,500 (at 7% per year) in interest alone. Then as you get all your debts paid down, you can invest in other areas.

There are so many variables in farming. For example: How much land is available or how much help? All these need to be taken into consideration as we plan our farm. As for labor, there is quite a bit of difference between a young couple starting on their own or a family that has several boys that need work and maybe several grandparents to help a bit. So especially if you are short on labor, think of using your time efficiently. How do I handle a good number of cows with a reasonable amount of labor? This is where the beauty of a grazing dairy comes in. Also the start-up costs are usually lower.

Another point is if you increase 10 and 20 cows but have to have more facilities, more equipment, and possibly hire labor on a limited amount of acres, you may not be all that much better off.

What is the most important part of a financially successful farm? In my opinion animal husbandry would have to rank number one. We will quote Proverbs 27:23: "Be thou diligent to know the state of thy flocks, and look well to thy herds." All the plans and number crunching will be irrelevant if this is not met.

I don't want to make it sound more difficult than it actually is. But I would strongly recommend that a young would-be farmer educate himself on this as much as possible. This can be done by spending time with an experienced dairyman and/or reading the various dairy publications that are available. Remember, the school of hard knocks can be very cruel and it is much easier to learn beforehand.

Just how important is this? Let's take a look at a possible scenario. Keep in mind that this is an example only and that these farmers are fictitious.

Farmer A and Farmer B both begin farming the same year. They both buy 35 cows at an average of $1,500, for an investment of $52,500.

Farmer A thinks cows are important and has learned his lessons well, and in five years he is averaging well over $1,000 in profit per cow per year. He has grown from his original 35 cows to 45 and has 35 heifers for a total of 80 head. He has increased the quality of his herd, and they are now worth an average of $1,800. His 80 head now have a worth of $144,000 for an increase of $91,500 on his herd alone.

Farmer B doesn't think cows are all that important. He is soon battling mastitis and Johne's, to name a few, and in five years he has declined to 30 cows and his net profit is under $1,000 per cow. He may have 25 heifers for a total of 55 head, but they have actually decreased in value.

These two examples may be a bit extreme but are very possible.

Another word of caution: Although we do need a plan to address our family's needs and our debts, let's not get too wrapped up in having more than the neighbors or whoever, lest we fall like the rich man we read of in Luke 12. Remember, God first, family second, and only then our farm.

Would I encourage young couples to begin farming? Absolutely. As I think back, yes, we struggled with inexperience and high debt those first years, but we had a vision of how things should be. And it has become even better than we could have imagined. It is truly rewarding to own a piece of God's creation. Or do we really own anything? Would not caretaker be a better word? Let us care well for our families and our farms. Let us cherish God, our families, our heritage, our freedom, and yes, our farms too.

Start-up Costs I

Note: The three scenarios in this section represent the middle to high side of start-up costs for Amish farmers. If you are a would-be farmer and want to begin on the proverbial shoestring, we urge you to find ways to do so. However, do so with the wisdom that knows that going the cheap route at times ends up being the most expensive. Understanding the principles that guide right stewardship will help you. Make yourself a student of good stewardship.

Getting started in dairy can be a costly venture, but there are also opportunities to be frugal. Let's consider the basic components of land, cows, buildings, and various types of equipment. The numbers shared are actual numbers from our experience, and are shared only as an example.

First is land. We purchased our farm 11 years before we began milking, and paid for it primarily with off-farm income. The $2,550 per acre price we paid for it then is unheard of in today's large community public real estate market. It was possible then because of the kind consideration of the seller who chose not to use the public auction method of sale with its multiple parcels. Fortunately, there are still parents who pass on the family farm to a son or daughter at a reasonable price. Other options to consider include renting land or purchasing land in a smaller community where it is more affordable.

Once you have land, you need cows. We purchased 40 yearling heifers in the spring at an average cost of $965, and another 8 bred heifers in the fall at an average cost of $1,165. This brought our total cost on 48 head to $47,920.

Every farmer needs some buildings. We were blessed with an older barn. We spent $6,200 to repair and improve it. Next to the barn we constructed a 22' x 42' milking parlor and milk house combination. Behind the parlor is a 22' x 39' holding area and a 30' x 75' concrete loafing area.

The materials for this complete milking complex totaled $45,830. Labor paid to construct this came to $5,790 in addition to much donated labor, provided by neighbors, church members, and other friends. We spent $1,619 to rent various machines and $6,410 for excavation, which included moving and packing a fair amount of dirt. We were able to house our heifers in the barn the first winter, but have plans to add a hoop building prior to the coming winter. We are not done spending on buildings. When planning a building project always allow for a 15% cushion above the estimated cost.

Our final consideration is equipment. Remember that equipment has the tendency to depreciate, while cows appreciate. But we still need to have at least some equipment. We spent $15,202 for milking equipment. That includes 8 milking units that swing, a 1,000-gallon bulk tank, an oil-less vacuum pump, and all the pipeline components. Then we shelled out another $5,804 for labor to install this equipment. These costs were the greatest surprise in our project. A simple vacuum system with bucket milkers could be purchased and installed for quite a bit less. For milking equipment installation, I would recommend a quote rather than an estimate. To power the milking equipment we purchased a new 50 HP Kubota diesel engine with a 30 KW generator, and installed them in a 10' x 16' mini barn. That separated us from another $12,577.

Other equipment purchases included a used 46 HP skid loader for $9,500 and a line of hay equipment for $5,100. Previously owned equipment included a tractor, feed grinder, mixer, and manure spreader, for which we had paid another $4,800 sometime during the past 11 years. Since we are graziers and do not raise corn, we did not need to purchase any equipment to work the soil, but may need to do so in the future for pasture renovation. We paid $1,600 for a team of Belgian geldings, which complement our older team of mares owned previously.

Start-up costs can vary depending on what you are starting with and where you want to go. I've shared our experience with land, cows, buildings, and equipment. Your experience will certainly be somewhat different. There are many factors involved. We could have saved money by installing the parlor in the barn, but we feel there are benefits of having it in a separate building. I was working away full-time during all of the construction. I'm sure if I would have quit my job, we could have saved money on the project, but there would have been lost wages. If your carpentry skills are good to excellent, you can build a parlor for far less than we did. It is important to do your homework before you buy. Be thorough and educate yourself on the details. Enlist and accept the help of friends. There is a large and growing group of farmers who are willing to be involved with the construction of dairy start-ups. Together we can accomplish more.

Start-up Costs II

By the time we were in our mid thirties, my wife and I had talked a lot about owning our own farm. Then the opportunity came to buy one in our neighborhood. At the silent auction of this farm we ended up being the final bidders. We ended up with 110 acres for $440,000. On our 110 acres we had two houses, a big barn and several outbuildings that were set up to milk a few cows by hand and cool the milk in the springhouse.

Here are some of our costs for dairy start-up in 2005. First we built a lean-to on the main barn for a parlor and milk house. The cost was $6,570. Then we tore out several box stalls in the main barn and built a 50-cow feed bunk and loafing room. That cost $2,728. Here's a list of other expenses: galvanized pipe and 4" casing to build parlor - $2,000; plumbing in milk house and parlor (including in-floor heat) - $1,460; 4 milking units, dumping station, in-line filter, reservoir jar, and milk hose - $2,600; used vacuum pump - $1,150; 700-gallon milk tank - $1,850; complete setup of 10-horse cooling compressor and lights throughout the barn - $7,385; battery fencer and related supplies - $1,500; purchase and setup of 37 HP diesel - $5,882. Into this whole setup we poured 118 yards of concrete for $8,614. All of this was done with very little labor cost. We had a lot of help from neighbors, friends, and relatives.

Now we had a setup to milk 50 cows. We bought 27 heifers and 21 cows. Since our payments began the first year, we wanted our milk production to start the first year too. Our heifers cost $1,490 each. Our adult cows cost $1,800 each.

Now we were ready to milk cows on our 110 acres, but other equipment was needed as well. We needed equipment to make hay, haul manure, and mow pasture. Additionally, we would need some cultivation equipment. Here's what we bought: 4 draft horses - $5,500; 2 sets harnesses and collars - $1,800; rebuilt hay mower - $550; New Holland roller bar hay rake - $700; used square baler with new 18 HP Honda motor - $1,715; hay wagon - $1,075; elevator - $400; new 110 Pequea manure spreader - $3,430; new heavy Pioneer fore cart - $580; skid loader with bucket and pallet forks - $9,900; Pioneer riding plow - $275; spike-tooth harrow - $90; double disk - $160; cultipacker - $125; eveners and neck yoke - $160.

Now in 2007 my wife and I and our eight children are enjoying the farm. We milk 53 cows and have the entire farm in grass.

Start-up Costs III

Our farming venture started when a group of 49 Holstein-Jersey cross yearling heifers were purchased from a local heifer grower for $800 each. The heifers were bred to freshen the following spring. Forty of the original 49 cows were kept for our seasonal herd.

By early spring approximately 65 acres vacant land was parceled off on an existing farm with an agreement to rent at a low cost until land would be purchased at a later date. Construction began shortly thereafter. $4,300 was spent for a 38' x 72' hoop building, with construction on living quarters beginning shortly thereafter and by early summer we were able to move into our newly built home. Work on the parlor and the milk room began that fall.

By the time the first milk hit the tank the following spring, we had spent $17,000 on infrastructure including parlor, milk room, engine room, holding pen, and some concrete around the parlor. Another $15,000 was spent on equipment which included bulk tank, cooling equipment, vacuum pump, bucket milkers, wash vats, and storage racks and an air compressor and air tanks for water supply. Equipment is run off a line shaft powered by a rebuilt diesel engine. Some water lines were run in the parlor floor for hot water heat. Costs for fence at start-up came to $4,500. Our goal was to minimize investing in assets that depreciate and focus instead on appreciating assets (cows). This is especially important in the start-up years. This we did by doing some outwintering and borrowing or renting machinery rather than owning it. In years of low milk prices we focused on making our farm payments and paying our living expenses, and in years of good milk prices improvements are made on the farm. In other words, after the basic essentials were established, the farm had to pay for its own improvements.

By the second calving period we had built another 38' x 100' hoop building on a five-foot pony wall for a bedded pack and freshening pens. That total came to $18,250. Another $14,200 was spent for a feeding area with a roof, concrete, and headlocks.

By the end of the second year, our line of equipment that we owned was a plow, disk, sickle bar mower, sprayer for liquid fertilizer, and fore cart with harness and eveners powered by three draft geldings, which came to a cost of $12,400. We do not own a skid loader, but rent it when one is needed.

By the end of the third year, we had spent, in addition to the above figures, $4,900 for parlor and milk room, $2,000 for milking and cooling equipment improvements, another $9,750 for fence, $2,500 for liquid manure storage, $5,000 on the hoop building, and $2,100 for cow lanes.

The drawbacks to starting out with bare bones is that there are days when conditions are less than ideal and it is not necessarily the easiest route, although we are now glad we took this approach. We had a background in conventional farming, so when we began our operation in grass-based farming it required a different way of thinking and a different way of doing things with a different setup. Had we built and bought everything before we began milking instead of building and buying as we saw the need, we would have buildings and machinery that would not be as well suited to grass-based dairy.

In conclusion, we are excited and thankful to God for the opportunities there are today in grass farming for young families working together in harmony with nature. May we use it to His glory.

Record Keeping

Like putting the hand to the plow (the first and most important step in seedbed preparation), a good record keeping system will keep you on track. It will help you stay profitable and will let you know exactly where you stand financially. Use the right tool for the right job. A farm account book is a must in making sound financial and other management decisions. Most people who struggle financially fail on the spending end, not on income.

Categorizing income and expenditure will quickly tell you where money is coming from and where you are spending it.

Records should be kept on a monthly basis and added up quarterly as well as annually.

Income and expenditure should be compared quarterly and annually. Categories can be compared and percentages figured to compare ups and downs in expenses. Keep your records current at all times.

The first week of November is a good time to do tax estimates. Consult your tax man for specifics in tax rules and depreciation schedules. This gives two months to shift expenses and get tax benefits.

Vision + Plan + Discipline

1. VISION: Where you want to go. Set realistic goals.
2. PLAN: The financial accounting system you are going to use.
3. DISCIPLINE: Doing it.

Discipline is the big culprit. Carry a pocket calender and jot down all important details, numbers, and everything that happens. Only two hours after something has been said or has happened, many people cannot recollect the important numbers, details, or thoughts concerning it.

A good record keeping system will serve you well through good years and lean ones as well.

Financial Pitfalls I

What are the financial pitfalls in farming? The answers to this question can vary from farm to farm or person to person depending on the resources, knowledge, or experience prior to starting up.

One financial pitfall is not to have an approved loan in place before you start. Find out approximately what your income should be by asking for information from other farmers that use the same method of farming as you plan to. Get estimates on buildings, setup, chattels, etc. Add at least 10% on top to cover unexpected costs. Take this information and find an approved loan before you start. Do not spend more than you absolutely have to in the beginning. Remember that you have to pay interest on every dollar you borrow. Keep your payments as low as possible, but flexible so you can pay more on the principal whenever possible. Maintain a savings account to cover unexpected costs.

Spending a lot of money on expensive setups can be a pitfall. For example, an old tire filled with concrete and a steel fence post in the center along with spring gates can serve as a portable fence in the barnyard or feed lot and is much cheaper than a permanent fence and pipe gates.

Parlors can be very expensive or inexpensive and still do the same job. Chains can be used in some places instead of pipe gates. Do your own work with the help of friends instead of hiring someone to do everything. Installing a vacuum line is easy enough. Build your own brackets from wood or whatever you have on hand. Pouring concrete when needed is another job that can be done with the help of fellow farmers and friends.

Another pitfall can be investing too much money in buildings and expensive equipment that depreciates. Some equipment may be practical to borrow from neighbors in the beginning, rather than buying something you use twice a year and the rest of the time it's sitting around rusting and gathering dust. Learn to fix your own

equipment instead of paying a mechanic high dollars to do the job.
Invest your money in cattle; they appreciate. Buy cattle from a healthy herd, rather than at the weekly livestock auction. Buying cattle from a herd with Johne's and mastitis is a very expensive mistake. Remember, they are your source of income.

Following the advice of every salesman that shows up is a pitfall. Remember salesmen are out to sell their products and you are the one to pay the bills. Remember the age-old saying, "A penny saved is a penny earned." If something is supposed to be "so good," check with several folks you know that used it before you buy.

Financial Pitfalls II

One of the things you can be careful of is buying horses and cows whose price is higher because they carry a good pedigree. Unless a person sells breeding stock, the higher price is rarely justified.

Be careful of purchasing fancy factory-made fence handles, reels, switches, volt meters, or spring gates. Many of these items can be homemade with little or no expense. For example, a blade of grass serves well as a fence tester.

Parlor equipment is another place to cut costs. Though it is good, stainless steel is expensive, and one could use a cheaper substitute to get started. Later, when your debt load eases, these items can be upgraded.

Chains and sliding boards save money on gates and can work well in some situations. They become more expensive, though, if they're not secure and your cows end up in the neighbor's cornfield, or worse yet, out on the road.

I would encourage buying used field equipment and loaders until you have eliminated some of your debt.

Attempting to save money by raising colts from your draft mares doesn't pay if you feed your hay and grain to the colts instead of to your replacement heifers.

Considering the price of corn versus milk, feeding grain must be carefully analyzed to make sure it is kept at a carefully balanced level. Yes, it produces more milk, but the question you need to ask is, does it produce enough increase to justify the grain cost?

Wrapping high moisture hay can be too expensive. If weather permits and you have room to store hay, put it up dry.

More equipment, labor, and of course money is required to raise small grains and corn versus grazing and making hay.

When selecting draft horses choose a horse that is built short and stocky. They need to eat less feed to stay in good condition.

Consider using old tractor tires with a concrete bottom as a water trough. They are much cheaper than concrete troughs. You can make your own frost-free trough as well.

Using bedded pack instead of free stalls saves money by requiring less equipment and labor, plus cows are more comfortable. The manure this produces is better for your fields than liquid manure.

Financial Pitfalls III

Here are financial pitfalls I have seen and think are worth the consideration of young start-up farmers.

When you borrow money, tell the whole truth about your financial condition. Don't try to cover financial problems that need to be talked about to a lender. When you borrow money from friends or relatives, you are expected to be entirely open about your finances. Not doing so causes lack of respect and trust, which are both needed to maintain a good relationship.

Be careful how much and to whom you discuss finances. Bragging about your last year's profit may be a pitfall for another who thinks, "This must be easy, so I'll try it too."

Till you can afford it, avoid buying the best machinery. Be diligent in maintaining well what you do have. Do the repairs yourself whenever possible.

Don't make hasty decisions with salesmen. Take time to sleep over it and ask counsel from your wife and someone with experience in the area in question.

Ask advice when things aren't going well. Yes, it's humbling, but that is good for us. Later we can look back on those days and appreciate the advice given to us.

There is a big difference between our needs and wants. Try to identify which one it is before considering the other options. Doing this early on saves you from becoming emotionally attached to a want for so long that the want transforms itself into a "need." Be careful to not constantly wish for something that is off limits.

Be content with what you have. You will be happier for it. Avoid jealousy, trying to be first or on top. Be optimistic. Think of others. When you do make mistakes, admit it.

Let's involve our children whenever we can. We may have to stay at home more when the children are small. Children want to be around Daddy too. We need to work hard, but there is more to life than work. Children can help with the garden. Teach your children to eat vegetables and home-cooked meals at a young age. Train your children to do with less. We don't always have to have ice cream or snacks.

A faithful man shall abound with blessings: but he that maketh haste to be rich shall not be innocent (Proverbs 28:20).

Mortgage Payment Calculator

To calculate your monthly mortgage payment from this chart:

1. Find the number of years for the term of your loan across the top.
2. Find the percentage rate of your loan along the left side.
3. The place on the chart where the year column meets the percentage row is your factor number.
4. To calculate your estimated monthly payment, multiply your factor number by the amount of your loan in thousands (for example, 40 for $40,000.00, 75 for $75,000.00, etc).

Here's an example: If you are financing $50,000.00 for 15 years at 7%, multiply 50 times 8.99 (the factor number). Your total is $449.50—so your monthly payment, including principal and interest, will be $449.50.

% Rate	Years	5	10	15	20	25	30	35	40
6		19.33	11.10	8.44	7.16	6.44	6.00	5.70	5.50
6.25		19.45	11.23	8.57	7.31	6.60	6.16	5.87	5.68
6.5		19.57	11.35	8.71	7.46	6.75	6.32	6.04	5.85
6.75		19.68	11.48	8.85	7.60	6.91	6.49	6.21	6.03
7		19.80	11.61	8.99	7.75	7.07	6.65	6.39	6.21
7.25		19.92	11.74	9.13	7.90	7.23	6.82	6.56	6.40
7.5		20.04	11.87	9.27	8.06	7.39	6.99	6.74	6.58
7.75		20.16	12.00	9.41	8.21	7.55	7.16	6.92	6.77
8		20.28	12.13	9.56	8.36	7.72	7.34	7.10	6.95
8.25		20.40	12.27	9.70	8.52	7.88	7.51	7.28	7.14
8.5		20.52	12.40	9.85	8.68	8.06	7.69	7.47	7.34
8.75		20.64	12.54	10.00	8.84	8.23	7.87	7.66	7.53
9		20.76	12.67	10.15	9.00	8.40	8.05	7.84	7.72
9.25		20.88	12.81	10.30	9.16	8.57	8.23	8.03	7.91
9.5		21.01	12.94	10.45	9.33	8.74	8.41	8.22	8.11
9.75		21.13	13.08	10.60	9.49	8.92	8.60	8.41	8.30
10		21.25	13.22	10.75	9.66	9.09	8.78	8.60	8.50
10.25		21.38	13.36	10.90	9.82	9.27	8.97	8.79	8.69
10.5		21.50	13.50	11.06	9.99	9.45	9.15	8.99	8.89
10.75		21.62	13.64	11.21	10.16	9.63	9.34	9.18	9.09
11		21.75	13.78	11.37	10.33	9.81	9.53	9.37	9.29
11.25		21.87	13.92	11.53	10.50	9.99	9.72	9.57	9.49
11.5		22.00	14.06	11.69	10.67	10.17	9.91	9.77	9.69
11.75		22.12	14.21	11.85	10.84	10.35	10.10	9.96	9.89
12		22.25	14.35	12.01	11.02	10.54	10.29	10.16	10.09
12.25		22.38	14.50	12.17	11.19	10.72	10.48	10.36	10.29
12.5		22.50	14.64	12.33	11.37	10.91	10.68	10.56	10.49

*Used by kind permission of Beishtand

Examples of Making Larger Monthly Payments

Compound Period .... : Monthly
Nominal Annual Rate.................. : 7.000%
Effective Annual Rate.................. : 7.229%
Periodic Rate................................ : 0.5833%
Daily Rate..................................... : 0.01918%

Example 1: *25 years—regular payments*

Cash Flow Data

Event	Start Date	Amount	Number Period	End Date
1 Loan	07/01/2005	160,000.00	1	
2 Payment	08/01/2005	1,130.85	300 Monthly	07/01/2030

	Payment	Interest	Principal	Balance
Grand Totals	339,255.00	179,255.00	160,000.00	0.00

Example 2: *Extra $50.00 per month*

Cash Flow Data

Event	Start Date	Amount	Number Period	End Date
1 Loan	07/01/2005	160,000.00	1	
2 Payment	08/01/2005	1,180.85	267 Monthly	10/01/2027
3 Payment	11/01/2027	1,932.62	1	

	Payment	Interest	Principal	Balance
Grand Totals	317,219.57	157,219.57	160,000.00	0.00

By paying $50 extra every month, you would save $22,036 and cut 2¾ years off this 25-year example.

Example 3: *Extra $100.00 per month*

Cash Flow Data

Event	Start Date	Amount	Number Period	End Date
1 Loan	07/01/2005	160,000.00	1	
2 Payment	08/01/2005	1,230.85	243 Monthly	10/01/2025
3 Payment	11/01/2027	1,397.90	1	

	Payment	Interest	Principal	Balance
Grand Totals	300,494.45	140,494.45	160,000.00	0.00

By paying $100 extra every month you would save $38,761 and cut 4¾ years off this 25-year example.

*Used by kind permission of Beishtand

Sources of Information

Salesmen

A salesman is a person employed to sell merchandise either in a territory or in a store. He is not usually a good source of information. Although he may seem to be concerned about your bottom line, he is much more willing to provide you with bragging rights such as pounds of milk per cow, tons per acre, bushels per acre, etc.

Don't buy on impulse from a salesman. If his product seems good, get his phone number and tell him you'll call him when you decide.

If the deal is good for only one day, get your wife involved or some other person that did not get the first hype, to look into it. It is not necessary to be rude to a salesman; simply give him very little of your time and none of your money, and he soon can't afford to stop at your place. Buying from a local dealer is a much better choice. Here you decide what you need, call in your order, and be done with it.

Consultants

The best consultants are model farmers who farm like you are attempting to. These farmers won't profit from you and will be truly concerned about your bottom line. You need to ask yourself several questions: Am I able to copy this model farmer with regards to farm size, level of knowledge and experience, financial backing, land fertility, family labor, and even ambition?

Pasture Walks

Pasture walks are often a good place to learn and to meet other farmers that are willing to share experience, advice, and knowledge that is hard to find any other place. Bring your notebook and write the different opinions down, then review them when you get home.

Grazing Conferences

Consider attending Farm Family Field Day, North Central Ohio Grazing Conference, and Northern Indiana Grazing Conference. There are many other good sources of information, but here again, keep in mind that the speakers are usually not in their first or second year, and especially if figures are thrown out, these figures are not meant to be achieved in the first few years. These speakers also come from a variety of climate soil types and marketing options. Nevertheless, be sure to attend these meetings to get your spirits renewed for the next year.

Periodical Publications

1. *Graze* Magazine is #1 on the list of information sources for dairy graziers. It is written by active farmers. Usually in-season articles, how-tos, and why we do what we do. Also Q & A section. Write to: *Graze,* P.O. Box 48, Belleville, WI 53508. One year is $30, two years $54.

2. *Farming* Magazine has good stories to read that gets you to appreciate nature and brings out the romantic part of farming. Helps you recognize the bird's song, the sunrise, and the sunset. Contains handy hints and recipes. Write to: *Farming,* P.O. Box 85, Mt. Hope, OH 44660. One year is $18, two years $32.

3. *The Stockman Grass Farmer.* A good grass farm magazine focused on beef cattle. Write to: *The Stockman Grass Farmer,* P.O. Box 2300, Ridgeland, MS 39158. One year is $32, two years $56.

4. *The Milkweed.* The best source of information on dairy prices. Predicts prices for milk, cheese, cows, heifers, and almost anything in the dairy world. Includes a weather map and availability of feed throughout the world. Write to: *The Milkweed,* P.O. Box 10, Brooklyn, WI 53521. One year is $45, two years $85.

5. *Hoard's Dairyman.* Listed as the national dairy farm magazine, but most of the information is for large dairies. Yet it will give you insight on what is going on with dairy business throughout. Lately they are starting to print some stories on grass farming. Write to *Hoard's Dairyman,* P.O. Box 801, Fort Atkinson, WI 43538.

Authors

Here are five authors whose books will provide insight and encouragement to farmers who are getting started. Allen Nation, Joel Salatin, Jim Gerish, Louis Bromfield, and Gene Logsdon.

The more I am acquainted with agricultural affairs, the better I am pleased with them. Insomuch as I can nowhere find so great satisfaction as in these innocent and useful pursuits. In indulging these feelings, I am led to reflect how much more delightful to an undebauched mind is the task of making improvements on the earth, than all the vain glory which can be acquired from ravaging it by the uninterrupted career of conquests. -*George Washington*

Grazing Herd Management

A new farmer has three important "teachers" to show him how to operate a farm. He would be wise to utilize all of the three. They are: your fellow farmers, reading material (e.g. books, newsletters, and magazines), and nature as God created it.

I believe the first two teachers and how to utilize them will be addressed by other writers on this project.

My goal is to bring your attention to the third teacher—nature. I want to tell you how you can learn to use nature to benefit and profit by on your farm.

In the beginning God created the whole earth and life thereon and He called it good. Though the Bible does not specifically say so, I believe He created the thorns and thistles at this time as well. I like to think that thorns and thistles were put in as safeguards. Fallen man is selfish, and if God through nature would not hinder the farming practices of agriculture that are out of sync with nature by thorns and thistles, mankind would farm himself to utter ruin. Proof is here that he is trying to do so with unsustainable farming methods currently.

So you want to start a dairy farm? Cows will be needed. Let nature be a guide in selecting your herd. Do not invest money in cows that need support or artificial inputs in order to stay alive and productive. If the cow you are buying needs more than twenty dollars of annual vet care, or if she needs corn silage and more than ten pounds of grain per day to remain productive, then you should not buy that cow. Find a cow that will survive and even thrive and be productive and reproductive on good pasture and a minimum amount of other feed. Oftentimes the cow you will want may not be producing what high-producing herds want. The cows you need can be found if you take time to seek them out.

We recommend a spring seasonal calving cow. No, it is not an absolute must. Please understand, though, for the sake of the cow's survival, nature will not allow cows to calve at other times of the year. (Here come the thorns and thistles again.) You will greatly complicate your operation by not having the cow calve in time with nature.

Learn to take good care of baby heifer calves. Feed them lots of whole milk and let them suck it out of nipples. Placing them in groups or barrel feeders allows you to feed and care for a lot of calves in a hurry. Feed the calves grain and hay and get them out on lush pasture as soon as possible. They should be on good pasture when they are weaned but should be fed free-choice hay and appropriate amounts of grain. Start rotating these calves to new and fresh pasture quite often and don't bring them back to the same pasture for at least a month to break the worm cycle. All the while, keep a close eye on them for worms, pinkeye, and coccidiosis. Graze these heifers as much as possible to build them into pasture-loving cows. As the twig is bent, the tree is inclined.

Cows want to be well fed. Do everything you can to keep them well fed on pastures, because that is your cheapest feed source by far. But if your cows cannot be well fed on pasture because it is winter or it is too dry or you have not managed right to have enough pasture for your cows, then you must feed them something else to satisfy them and keep them productive. Do this just because you have to, not because you want to. Hay or balage is probably the best choice replacement for pasture when grass is in short supply. Have good, efficient milking setups and an easy way to feed cows in the barn or bunk. This does not mean it has to be an expensive setup.

Sometimes small changes make big differences in time spent chor-
ing.

Let the cow do most of the work. What you don't arrange to have her do, you or some machine will have to do for her. God has equipped the cow to be able to harvest most of her feed herself by grazing. She should "haul" most of her own manure. She should be able to fight off most diseases and illnesses all by herself. She can do all these things if you provide good grass for her to eat from early April to late November.

Farmers spend a lot of time, thought, and money in the grass varieties that should be grown. It is not that important actually. What is important is finding grass varieties that stay in place for a long time. So the seeds you want to buy are probably going to be orchard grass, rye grass, and varieties that perform similar to these.

By far the most important pasture species is going to be white clover. All of your pasture should have white clover. Clover should be an indicator plant. In other words, if you cannot maintain clover in your grass swards, then something is wrong either with the soil or the way you are managing your pastures. Ask an experienced grazier to help figure out what is wrong when clover is missing from your pasture.

In nature, if something wants to die it deserves to die. Don't fret over some grass seed that goes away after three to five years from planting, no matter how well it did the first year or two. Your goal is to have better swards of grass after six years from seeding, not worse.

The best farmers pay a lot of attention to the soil. Keep in mind that there is a living community of earthworms, microbes, bacteria and fungi in the top six inches of your soil. When you have good living soil, this community of tiny critters is said to add up in total weight greater than the livestock feeding on the grass on top! These tiny, creeping things are busy working for you 24 hours a day, therefore you must take care of them. They like to be in the dark, so make sure they don't see the light of day very often by plowing. How they hate that! Think of the plow to be used in grass farming to cover a past mistake. Another way of keeping it dark for those creeping, crawling workers is to keep the grass swards thick enough and tall enough that the sun never shines onto any soil at all, but only on the grass and clover plants on top. Sunlight on bare soil will only do it good in very early spring when the air is cold and the creeping things are still in dormancy. This can warm up the soil faster so that grass can start growing sooner. Otherwise, nature hates having the sunshine hit the soil. Bare soil to nature is like skin removed from the human body. Just like a scab appears on removed skin, so does nature provide weeds to appear as a scab to cover its wound.

These underground critters also like to be fed on a regular basis. The cow herd can do a lot of this in the process of grazing. They will drop nutrients in the form of manure and urine and also tread in dying and extra grass. This is the type of food these critters love and thrive on. If they are not fed on a regular basis, many will starve and die. That is why you should always spread some manure on fields you have made hay off of. I have yet to see a hay baler make a pile of manure.

Feed the soil when nature feeds the soil. Nature gives soil a feeding in early summer when the grasses form seed heads and turn brown and die and go back to the soil. It also feeds it in October when the trees shed their leaves. So I think the best time of year to spread manure and soil amendments is from early summer to mid fall. Feed it in small amounts, often, rather than large amounts far apart. Just like we humans cannot eat all day for one day and then work for a week without eating anything, these critters like having a constant supply of food at hand.

Soil life likes open and airy profiles. Care should be taken to not compact the soil. Do not put the cows in wet areas for long periods of time. Plants with long taproots, such as dandelions or alfalfa are good plants to open the soil.

Wild animals in nature hardly ever get sick. That is because they have access to a large variety of plants and they select the ones to eat to keep them healthy. Give your cow herd access to fencerows and brushy areas where they can pick up browse such as elderberry leaves, raspberry leaves, blackberry leaves, plus other species of plants as well.

While nature is a great teacher and we must model our farms to mirror nature, we must also realize nature is slow. It often takes years to bring a farm into good production by natural ways. Sometimes a farmer does not have years of time before the banker wants his money back. Then you must use shortcuts. You may even have to buy and use unnatural things such as chemical fertilizers and poisonous sprays to get the type of pasture you want the first year or two. You might have to feed your cows more stored and costly feed to get them to produce enough milk to make ends meet. But always consider these things as temporary and not permanent solutions.

Very likely you will be selling your milk to someone who does not care how natural or unnatural you produce your milk. Perhaps they will pay you less for springtime milk. Or sometimes the customer wants a different type of product than what is easily done with nature's methods. Then you must change your farming methods to meet this demand. Just make sure you do not violate the laws of nature too greatly to do something for market's sake.

So then if you build your farm from the grass up and listen to the three teachers, your fellow farmers, good literature, and nature, then naturally good things should start happening for you.

I'm a great believer in luck; the harder I work, the more I have of it (Thomas Jefferson).

Be diligent to know the state of thy flocks, and look well to thy herds (Proverbs 27:23).

Choice of Markets

Choice of Markets I

Now that you're interested in dairying, you will need to choose a market for your milk. The direction you take depends on what your interests, resources, and experiences have been up to this point. It has to be profitable, enjoyable, and satisfying to you and your family.

The grazing dairy has a choice of markets that fit in well with the grazing system.

The **conventional market** is one option. It is the easiest to get into. The conventional market allows the use of chemical fertilizers, insecticides, pesticides, and herbicides when needed. It also allows you to use antibiotics when needed. It affords lower grain prices and more options on seeds. A more fluctuating pay price can be expected.

If you are willing to farm without chemicals and antibiotics, the **organic market** might be something you want to look into as you will receive a premium for your milk. It is, however, important not just to look at the pay price and ignore other factors such as higher feed, fertilizer, and seed prices. Some time must be spent keeping records of field crops (inputs and harvest), animal health records, etc. Also, you have a once-a-year certification fee and organic inspection to go through. You will need to learn about prevention in livestock, rather than using drugs for health issues. A buffer zone between conventional and organic fields needs to be maintained. You also need three years of transition to convert your farm to organic.

In the third year of the three-year transition period you must feed 100% organic feed but receive the lower conventional milk price. This third year can be difficult financially, but it can also be educational. The organic milk market tends to be more stable than conventional.

Another market potential in some areas is the **grass-fed market.** This market also fits well with the grazing dairy. It is easier for the start-up farmer in that it requires less paperwork and carries fewer restrictions on insecticides, pesticides, and animal medications. The grass-fed market often offers better milk prices than conventional, but its market tends to be more seasonal.

Choice of Markets II

So you're starting to farm? A noble profession indeed. A profession that surpasses the medical profession, for without farmers supplying the daily food for multitudes, our nation would starve, whereas the medical profession is geared to help the sick and wounded. Yes, we need medical help at times, but food for the multitude is needed daily!

I have several questions for you before you decide which milk market is the best option for you. What are you doing in your spare time right now? Are your hobbies related to agriculture? Are you helping farmers and getting more farming knowledge? What is the lay of your land? Is it hilly, rolling, or flat? Is your land chemically addicted? Is it very fertile or infertile? Is it well drained or poorly drained? What do you see when you walk over your farm? Do you see lots of earthworms? Are the cow patties disintegrating with the help of small insects or are they drying into buffalo chips? Are there lots of bare spots in the grass or do you see dense swards of grass? Are there areas with some grass and lots of hard-stemmed, acid-loving weeds? What is your soil pH? How much organic matter do you have?

What is your experience with cows? Do you think of them as beautiful scenery along the road? Have you helped a farmer milk several times or worked with cows for years? Do you know what to look for in healthy cows (e.g. cud chewing, eyes and ears alert, smooth hair)? Do you know how unhealthy cows look (e.g. cold ears, no appetite, rough hair)? Can you tell if a cow is in heat? Do you know the different stages of mastitis? Or the stages of freshening?

Do you have a passion for farming or is it just wishful thinking? Do your wife and family share your vision?

Be sure you and your wife see eye to eye before you give farming a try!!

A true wife is a harbor in the storms of life; a bad wife is a storm in the harbor.

Beside every successful man stands a faithful wife!

Here are two thoughts for the single wanna-be farmer: Of all the "investments" you make in life, be sure to "invest" in a faithful wife! Never marry for money. You can borrow it cheaper.

Are you enthusiastic about being a farmer or just looking for an exemption from being a day laborer? Enthusiasm is contagious... and so is the lack of it. Planning your work is good; doing it is better! Whatever you do, do it with all your might. Things done by halves are never done right. Farmers don't need to work hard—they need to work smart!

What do you do when the going gets tough? Do you face or flee problems? Do you seek solutions or make excuses? Do you ask questions to increase your knowledge or are you afraid to expose your ignorance? The only dumb question is the one that was never asked. Do you think this too shall pass and keep going? Tough times never last; tough people do! When the going gets tough, the tough keep going.

I detect you are all excited about twenty-five- to thirty-dollar organic milk. So now where do you score with your experience? Here are five levels of experience followed by suggestions that will help you.

1. ***No experience.*** Work for or donate some time to a successful farmer. Learn all you can before making any investment. This will help you from getting ahead of your headlights...and being in the dark!

2. ***Minor experience.*** Start at Grade B level.

3. ***Some experience.*** Grade A optional but not organic.

4. ***Several years hands on experience.*** Organic (optional), grass-based (feasible).

5. ***Lifelong experience.*** You grew up on the family farm; you have a farm that is alive with earthworms and dense swards of grass, not chemically addicted with high fertility and a soil pH between 6.5 to 6.8; you have well-drained soil with high organic matter; your family shares your vision—organic farming is an attainable goal.

If your experience level is one of the first three, ask lots of questions and work with experienced farmers to get hands-on experience. Learn from other people's mistakes—you won't live long enough to make them all yourself! Learn to ask the right question, to the right person, at the right time.

If you score below four, learn to swim in shallow water. Remember, Rome wasn't built in one day. Don't dive into deep waters before you can swim, for you could drown before you learn how to swim. You can always upgrade as your knowledge grows. If you start at the top, there is just one way to go—down. Farming is like the ocean—we have high tides and low tides. We need to ride the tides.

A young farmer needs to realize it takes several years to come to the level of success where the experienced farmer is at.

If you score four or five and plan to go organic, be sure to have a written agreement with your marketing company **before** you start your last transition year on 100% organic feed.

Never set up your loan or payments on organic prices. Set them up at Grade B prices, then if unexpected things happen, you will survive. If things go well, you will thrive!

Let's look at the different markets.

Grade B—Grade B milk is used for making cheese, and is slightly more tolerant on milk house sanitation, with once-a-year inspection. There is no PI testing, but Somatic Cell Count (SCC) is tested monthly. Bacteria is tested once a month as well. Maximum bacteria is the same for grades A & B. Maximum SCC is nearly same for grades A & B (750,000). Going over the maximum can result in being shut off.

Grade A—Grade A milk is used for bottled milk, ice cream and butter, etc. with tougher sanitation rules and biannual inspection. Grade A is subject to state and federal survey inspection. PI, SCC, and bacteria testing are done several times a month, then the results are averaged.

Grass-Based—Grass-based is a new market that is developing, with the same sanitation rules as Grade A. This market requires that no fermented corn silage or wrapped forages be used, only grass and dry hay. This market will require excellent grazing and quality haymaking management. A potential for the seasoned grazier!

Organic—Sanitation requirements are the same as Grade A. SCC requirements are some tighter than 400,000 but maximum SCC is same as Grade A. LPC testing is required.

This market requires that no chemical sprays, fertilizers, or treated seeds are used in the fields. Additionally, no drugs, hormones, or chemical sprays can be used on the animals. The organic farmer is interviewed by the organic inspector every year. The quality program for SCC, PI, and bacteria bring higher rewards than any other market, but reductions are taken off if your quality falls below standard. This market offers the highest prices.

Organic farming is more family friendly than chemical farming, but requires timely management to stay ahead of or to avoid problems, because those chemical and drug Band-Aids are not acceptable!

It should not be possible to identify organic cows from other cows (unless of course they look better). Take care of your cows and feed them a balanced diet. What is the gain of getting an organic price if cows are undernourished and give a poor amount of milk? You can't starve a cow to profit.

A thought on any market: If you are a free spender, there is no milk market that will support you. You have to be committed to farming and make sound investments. Sound investments create an honest return, sometimes manifold returns. Be aware of get-rich-quick schemes; they are not sound investments.

- If it sounds too good to be true, it probably is...
- Oftentimes medium prices are the best buy. Cheap isn't always cheap and expensive isn't always best.
- Be aware of small expenses; a small leak can sink a large ship. *-Ben Franklin*
- The definition of free spending: money used extravagantly and wastefully; money that brings no return whatsoever.
- He who spends before he thrives, will surely beg before he dies.
- Increased earning creates increased yearning.
- It's not what you get, but what you have left.
- If your outgo exceeds your income, then your upkeep will be your downfall.
- Some people don't start saving for a rainy day until it starts pouring.
- A barn can pay for a house, but a house won't pay for a barn!
- It's not the high cost of living, but the cost of high living.
- If money gets tight, do you buckle down to survive? Or do you try a new credit card?
- The luxuries of one generation are the necessities of the next.
- We make a living by what we get; we make a life by what we give.
- Many of life's failures are people who didn't realize how close they were to success when they gave up. *-Thomas Edison*
- The worst time to invest in new machinery is when the milk price is high. Farmers who set up loans at such a time are the first to complain when the price drops. *-Bob Pendleberry, milk inspector*
- Don't mind criticism. If it's untrue, disregard it. If it's unfair, keep from irritation. If it's ignorant, smile. If it's justified, learn from it!

I wish you the best in your farming venture. GO FOR IT!!!

Choice of Markets III

When considering your options for marketing milk, begin by making a list of all the options. Normally you have more choices in the fall than in the spring. Be sure to ask advice of fellow farmers in your area. First you may list the marketing option, then beside it list the benefits you can expect. Benefits could include price of milk, premiums, and market demand. Next to that, list the requirements of each option. Make sure it is within your power to meet those requirements according to your experience and management skills and in regards to your acreage and facilities. Be cautious and take plenty of time to thoroughly understand all the terms of a contract before signing a contract.

You will want to set achievable goals in writing for years one, two, three, five, and ten. Have a clear vision of your farming goals. And follow a plan as much as possible.

Maintain a good relationship between yourself and your milk company and hauler.

Make yourself a student of the cost per hundredweight that it takes to produce milk on your farm. As the manager of the farm, the cost per hundredweight is largely in your control. Constantly ask yourself: What are we doing? Why are we doing it? And then try to keep it all simple.

Be patient therefore, brethren, unto the coming of the Lord. Behold, the husbandman waiteth for the precious fruit of the earth, and hath long patience for it, until he receive the early and latter rain (James 5:7).